FLORA OF TROPICAL EAST AFRICA

BEGONIACEAE

VANESSA PLANA, MARTIN J.S. SANDS & HENK J. BEENTJE[1]

Herbs, subshrubs or sometimes shrubs, monoecious or very rarely dioecious, usually perennial, terrestrial or epiphytic. Stems herbaceous, sometimes succulent, woody, erect, reclinate, rhizomatous or tuberous (then acaulescent or short-stemmed), rarely a liane or climbing with adventitious roots. Leaves whorled in acaulescent species, otherwise usually alternate, more rarely distichous or opposite, stipulate (stipules persistent or caducous), petiolate, sometimes subsessile, occasionally peltate, usually asymmetric, the broader side often with enlarged basal lobe, but sometimes symmetric, entire to deeply dissected or palmately compound; glabrous or with simple or stellate hairs, sometimes with scaly trichomes; venation palmate, palmate-pinnate or pinnate. Leaf-axil bulbils sometimes present. Inflorescence axillary or terminal, androgynous or unisexual, usually cymose but sometimes racemose or compound, protandrous or protogynous. Bracts persistent or not; bracteoles often present (but rare in African species). Flowers usually white or pink, unisexual, tepals in 1 or 2 (*Hillebrandia*) distinct whorls. Male flowers with 2 or (3–)4(–10) free or variably fused tepals; stamens 3–many in an actinomorphic or zygomorphic cluster; filaments free or variably fused into a column; anthers always yellow (in our area), 2-celled, opening lengthwise or by short apical slits or pores, the connective sometimes extended (rare in Africa). Female flowers with 2–6(–10) often unequal tepals, free or partially fused; ovary inferior, or semi-inferior in *Hillebrandia*, broadly obovoid to globose or fusiform, with (1–)3(–7) wings or horns, more rarely wingless, (1–)2–3(–6)-locular; placentation axillary or occasionally parietal or septal, in cross-section the placentas entire or with 2(–4) branches, rarely the inner-facing surfaces devoid of ovules (as in sect. *Squamibegonia*); styles (2–)3–4(–7), persistent or caducous, free or partly connate, usually forked, sometimes more than once, the stigmatic tissue usually in a continuous spiral band, sometimes kidney-shaped or rarely occurring all over the style (in our area in sect. *Squamibegonia*). Fruit erect, pendulous or nodding, most commonly a capsule, the wings (or horns) subequal, unequal or differing in size and shape, more rarely (but fairly frequently in Africa) baccate or fleshy; usually dehiscent, usually loculicidally (or in *Hillebrandia* in between the styles), more rarely (especially in species with wingless fruits) indehiscent or dehiscing irregularly. Seeds numerous, minute, with collar cells, little or no endosperm and reticulate testa.

Two genera with more than 1500 species, all but one (*Hillebrandia sandwicensis*) belonging to the genus *Begonia*. Found throughout tropical and subtropical regions worldwide, with the exception of tropical Australia and the Pacific region from Fiji to the Galapagos islands.

[1] Vanessa Plana – Royal Botanic Garden, Edinburgh – spp. 1–8 & *B. horticola*.
Martin J.S. Sands & Henk J. Beentje – Royal Botanic Gardens, Kew – family and genus description, keys, spp. 9–19.

BEGONIA

L., Sp. Pl. 2: 1056 (1753); Irmsch. in E.& P. Pf. ed. 2, 21: 548–588 (1925) & in E.J. 81: 106–188 (1961)

Description as for the family, except for those characters distinguishing *Hillebrandia*, namely: perianth segments (10 in male flowers, 8–10 in female flowers) in two distinct whorls, ovary semi-inferior, and fruit dehiscing between the styles.

More than 1500 species; distribution as for the family except for Hawaii.

The genus *Begonia* has been divided into more than 60 Sections of which 16 are represented in Africa. Species belonging to five of these occur in the FTEA area together with *B. humilis*, a naturalised species of the American Section, *Doratometra*. A key to the Sections based only on species found in the FTEA is given on p. 21.

CULTIVATED SPECIES

Begonia heracleifolia Schlecht. & Cham. Perennial herb; leaves palmatilobed; flowers large, pink; fruit winged. Native to Mexico.
Kenya, Nairobi, *Grahame Bell* H310/58!; Tanzania, Amani, *Greenway* 1687!

B. glauca (Klotzsch) A.DC. Scandent fleshy herb. Native to Peru.
Kenya, Mombasa, *Jex-Blake* in EA 11094!

B. × *ricinifolia* A.Dietr. Perennial herb; leaves bronze; flowers pink.
Kenya, Nairobi, *Jex-Blake* in EA 11142!

Other popular cultivated begonias (species or hybrids) are likely to be found in the Flora area.
U.O.P.Z.: 145–146 (1949) mentions as cultivated on Zanzibar: *B.* × *argenteo-guttata* Lemoine (no specimens seen); *B. coccinea* Hook. from Brazil (no specimens seen); *B. cucullata* Willd. (as *B. semperflorens* Link & Otto) from Brazil (no specimens seen).

1. Leaves deeply lobed, or, if not lobed, with 5 acute to caudate points; ♂ flowers with 2(–4) tepals 5–14 × 4–16 mm; ♀ tepals 2(–4) similar to those of ♂; fruit without wings, 2–3 cm long .... 2. *B. oxyloba*
   Leaves not lobed (or shallowly lobed in *B. princeae* & *B. tayloriana*) 2
2. Petiole near junction with lamina with conspicuous ring of hairs or fimbriate trichomes (inconspicuous in *B. stolzii*); Fig. 1.1 3
   Petiole glabrous, uniformly hairy or densely tomentose near lamina, but without conspicuous ring of hairs or fimbriate trichomes 8
3. Plant to 13 cm tall, the stem covered with persistent stipules 12–30 × 23–50 mm; **T** 6, 7 .... 11. *B. schliebenii*
   Plant usually much larger; stem not covered in stipules 4
4. Leaves 1–2 × as long as wide; outer tepals of ♂ flowers 17–33 × 17–30 mm; tepals of ♀ flowers 10–17 × 7–15 mm; fruit 15–28 mm long .... 12. *B. wollastonii*
   Leaves 2–3 × as long as wide; tepals of ♂ flowers 6–19 × 4–15 mm; tepals of ♀ flowers 5–17 × 3–11 mm 5
5. Stipules fimbriate; flowers pink or orange 6
   Stipules entire; flowers pink or white 7
6. Stipules dentate-fimbriate, 3–15 mm; flowers orange .. 16. *B. sutherlandii*
   Stipules 2.5–4 mm, slightly fimbriate; flowers pink; **T** 7, only known from two collections .... 15. *B. stolzii*

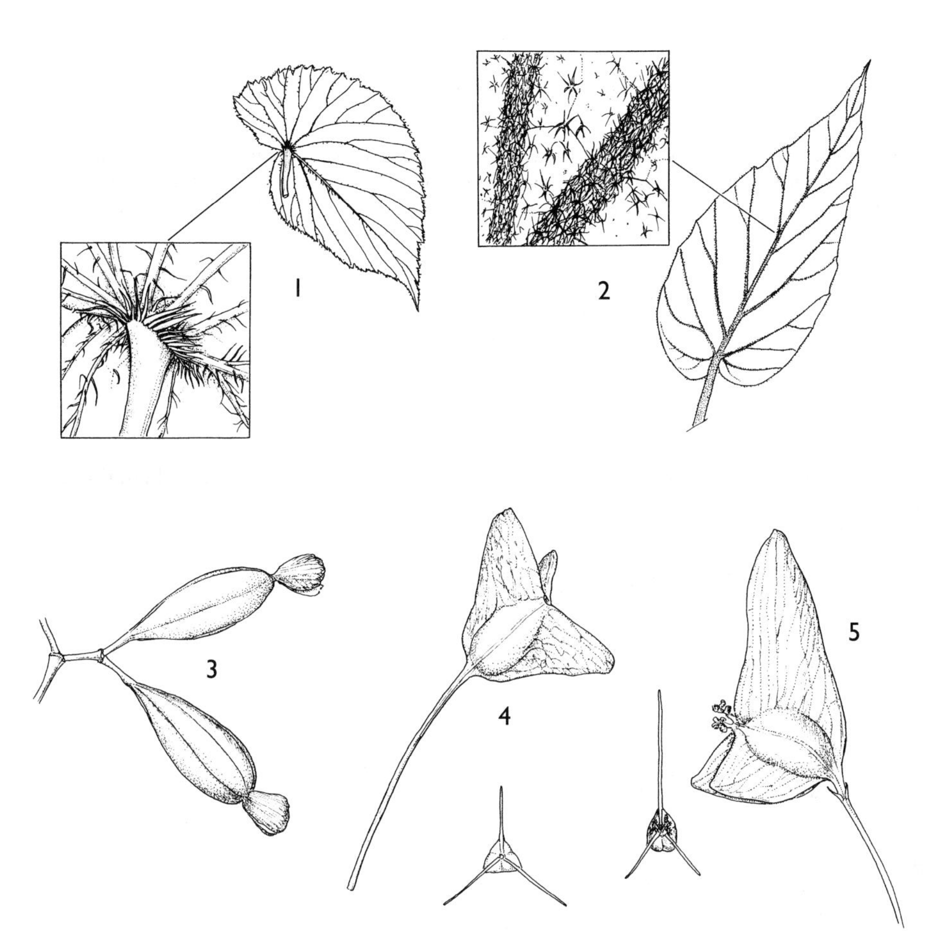

FIG. 1. *BEGONIA* key characters — **1**, petiole with ring of hairs or fimbriate trichomes, × ²⁄₃ (detail × 3); (with key lead 2); **2**, peltate scales with fimbriate margins, × ¹⁄₃ (detail enlarged); (with key lead 8); **3**, ovary/fruit, unwinged, × 1; **4**, ovary/fruit with equal wings, × 1; **5**, ovary/fruit with unequal wings, × 1. 1 from *Begonia wollastonii, A.S. Thomas* 517. 2 from *B. eminii, Poulsen et al.* 683. 3 from *B. oxyloba, Faden* 70/40. 4 from *B. sutherlandii, Polhill & Paulo* 1629. 5 from *B. johnstonii, Bally* 8038. Drawn by Juliet Williamson.

7. Leaves double-serrate; stipules 12–15 mm long, not markedly brown when dry; fruit 12–23 mm long, the wings subequal, rounded; **T** 3 only . . . . 9. *B. engleri*
   Leaves crenate; stipules 3–8 mm long, often dark brown when dry; fruit 7–11 mm long, one wing larger than others and angular . . . . 10. *B. johnstonii*
8. Stems and leaves with indument of peltate scales with fimbriate margins, resembling stellate hairs 9
   Stems and leaves glabrous or with simple hairs 15
9. Large bracts enveloping inflorescence; both ♀ and ♂ flowers with 2 tepals 12–43 × 10–40 mm, fused basally to form a perianth cylinder; fruits obovoid or globose 10
   Inflorescence not enveloped by large bracts; both ♀ and ♂ flowers with 4 free tepals; fruit fusiform 11

10. Leaf-base cordate; petiole 5–23 cm long .... 3. *B. ampla*
    Leaf-base cuneate; petiole 1–5.5 cm long .... 4. *B. poculifera*
11. Inflorescence unisexual 12
    Inflorescence bisexual, usually with both male and female flowers present 14
12. Stamens 20–40, the anthers ± 3 mm long .... *B. horticola* (p. 20)
    Stamens 7–19, anthers 0.8–2 mm long 13
13. Leaves ± symmetric, hairy on the veins beneath .... 5. *B. eminii*
    Leaves asymmetric, ± glabrous .... 6. *B. kisuluana*
14. Leaves asymmetric and falcate; **T** 3 (Usambara Mts) 8. *B. zimmermannii*
    Leaves almost symmetric; **U** 4 .... 7. *B. tatoniana*
15. Climber 1–10 m high or long; petiole tomentose distally; inflorescence unisexual; fruit unwinged, warty .... 1. *B. meyeri-johannis*
    Herbs to 60 cm high; petiole glabrous; inflorescence bisexual; fruit winged 16
16. Flowers orange; ♀ tepals 6–17 × 5–11 mm; ♂ tepals 7–18 × 6–15 mm .... 16. *B. sutherlandii*
    Flowers pink or white 17
17. ♀ tepals three, 10 × 7 mm; stipules 9–16 mm long; ♂ tepals 10 × 10 mm; **T** 4 (known only from the type) 17. *B. tayloriana*
    ♀ tepals five, 4.5–12 × 2–6 mm 18
18. Stipules 5–22 mm long; outer ♂ tepals 9–15 × 8–14 mm, ♀ tepals 6–12 × 4–6 mm .... 13. *B. princeae*
    Stipules 2–9 mm long; ♂ tepals 6–9 × 4–9 mm, ♀ tepals 4.5–8 × 2–5 mm 19
19. Ovary/fruit wings subequal 20
    Ovary/fruit wings unequal, one much larger ...... 18. *B. wakefieldii*
20. Leaves broadly ovate to suborbicular, with obtuse apex; **T** 7, 8 .... 14. *B. riparia*
    Leaves narrowly ovate, with acute apex; introduced .. 19. *B. humilis*

1. **Begonia meyeri-johannis** *Engl.* in Abh. Preuss. Akad. Wiss. 2: 305 (1892) & in Abh. Königl. Akad. Wiss. Berlin 2: 305 (1892); T.T.C.L.: 69 (1949); Wilczek in F.C.B., Begoniaceae: 14, fig. 2 (1969); Cribb & Leedal, Mountain Fl. S Tanzania: 53, pl. 7b (1982); Troupin, Fl. Pl. Lign. Rwanda: 166, fig. 56 (1982); Blundell, Wild Fl. East Afr.: t. 48 (1987); K.T.S.L.: 120, fig. (1994); U.K.W.F. ed. 2: 92 (1994); Klazenga et al. in B.J.B.B. 63: 278, fig. 4, map (1994). Type: Tanzania, Kilimanjaro, Mue R., *Meyer* 306 (B, holo.)

Herbaceous or woody climber, 1–10 m long, the flowering stems often hanging down, rarely epiphytic; stems often pink or reddish, rooting at the nodes, becoming woody with age; indumentum of simple hairs. Stipules narrowly triangular, 1–3 × 0.4–0.8 cm, apex acuminate, glabrous or with hairs at the base, caducous. Leaves green to reddish, thin to slightly fleshy, obliquely ovate, 4–13 × 2–8.8 cm, base obliquely cordate, margin denticulate to serrulate, larger teeth interspersed with numerous smaller teeth, apex acuminate; venation palmate; glabrescent above and below, venation tomentose beneath; petiole usually red, 2–12 cm long, junction with the lamina densely tomentose. Inflorescence axillary, unisexual or rarely bisexual, a dichasial cyme, many-flowered; bracteoles absent. Male inflorescence with 4–15 flowers; peduncle red or pink, 0.5–6 cm long; bracts weakly boat-shaped, elliptic to ovate, 1.3–2.5 cm long, chartaceous, with conspicuous veins, caducous; pedicels 1.1–2.6 cm long. Male flowers: tepals 4, white to pink, the outer suborbicular to broadly elliptic, 1–1.9 × 1.1–1.4 cm, the inner narrowly obovate, 0.8–2.6 × 0.2–0.4 cm; stamens yellow, many, ± 100, filaments free, anthers elliptic to shortly oblong, 0.4–0.8 mm long. Female inflorescence with 2–7 flowers; peduncle 0.7–3 cm long; bracts similar to those in male inflorescence, caducous. Female flowers: tepals 2–4, white to pink, the outer obovate to nearly orbicular, 1.7–3.3 ×

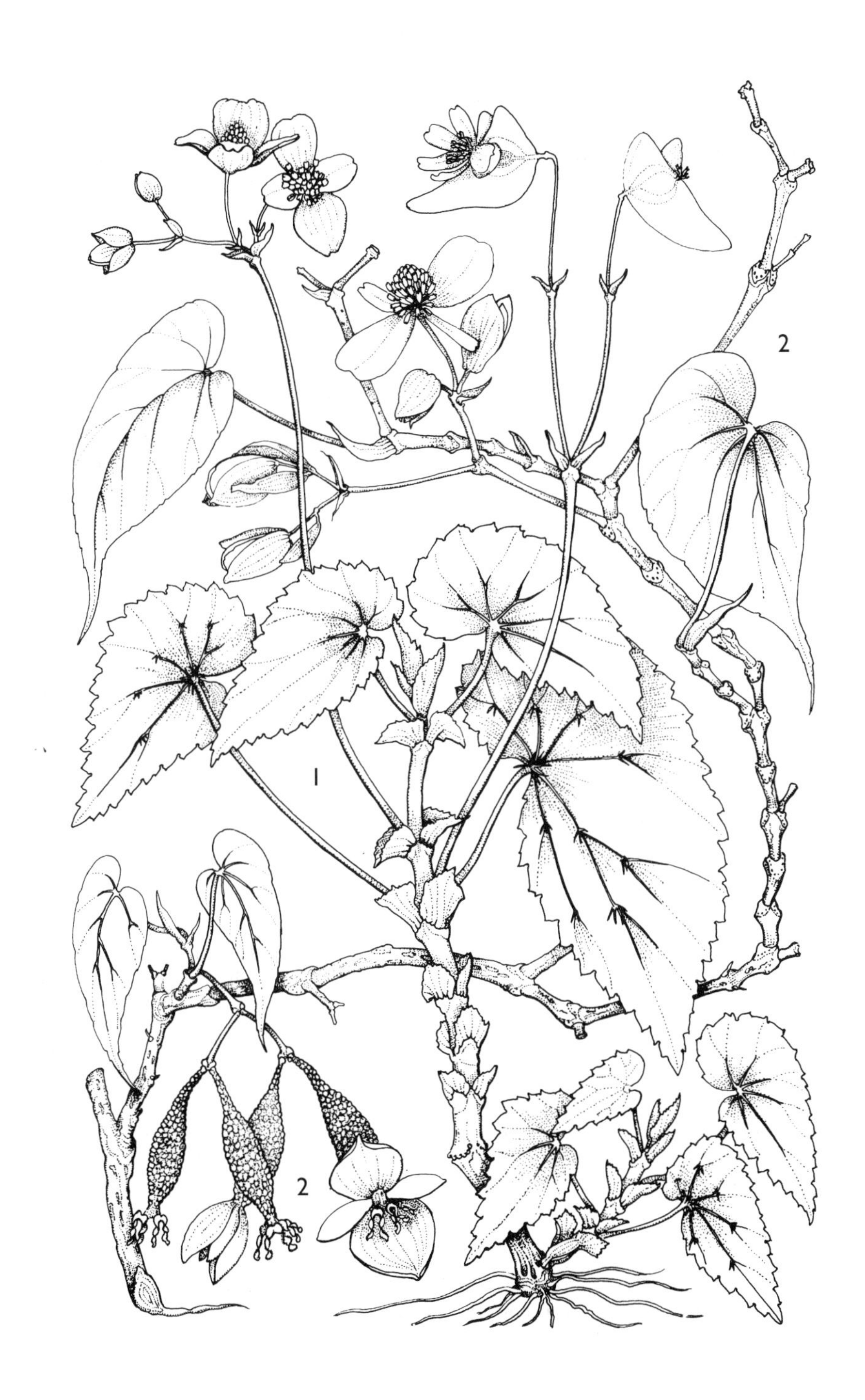

FIG. 2. *BEGONIA JOHNSTONII* — **1**, habit. *BEGONIA MEYERI-JOHANNIS* — **2**, habit. From Cribb & Leedal, Mountain Fl. S Tanzania. Drawn by Mair Swann, and reproduced with permission.

1.4–3.5 cm, the inner 1–1.7 × 0.4–0.6 cm, caducous or absent; styles 5, forked at the apex, spirally twisted twice, glabrous; stigmatic band yellow; ovary ellipsoid, 5-locular, without wings, surface verrucose; placentation parietal. Mature fruits greenish with white warts, fleshy, ellipsoid, 1–3 cm long, 0.8–1 cm in diameter, up to 4.5 cm including the extended stipe, terete in cross section, with narrow longitudinal furrows, verrucose (or warty), indehiscent, with persistent styles. Fig. 2, p. 5.

UGANDA. Toro District: Mahoma Valley, near Nyinabitaba [Nyabitabu], 10 July 1952, *Ross* 506! & 5 km W of Kilembe, 3 June 1970, *Lye & Katende* 5523!; Kigezi District: Kumba Swamp, 12 Feb. 1945, *Greenway & Eggeling* 7110!

KENYA. Kiambu District: Karamenu River Valley, 18 Aug. 1974, *Faden et al.* 74/1329!; Mt Kenya, Kamweti track at crossing of Thiba R., 10 Aug. 1985, *Beentje* 2196!; Kericho District: Chemasingi Tea Estate, 7 Dec. 1967, *Perdue & Kibuwa* 9298!

TANZANIA. Kilimanjaro, Marangu, 22 May 1961, *Machangu* 73!; Kilosa District: Ukaguru Mts, Mt Mnyera, 1 June 1978, *Thulin & Mhoro* 2833!; Njombe District: Livingstone Mts, 1.5 km W of church at Madehani, 13 June 1992, *Gereau & Kayombo* 4699!

DISTR. **U** 2; **K** 3–5; **T** 2, 3, 6, 7; Congo-Kinshasa, Rwanda, Burundi, Malawi

HAB. Moist forest, swamp forest, may be locally common; 1350–2800 m

CONSERVATION NOTES. Least concern (LC); widespread

NOTE. Flowers are often described as being fragrant, an unusual characteristic in African *Begonia*. It is strange that this plant does not seem to occur on Mt Elgon.

2. **Begonia oxyloba** *Hook.f.* in Oliv., F.T.A. 2: 573 (1871); Wilczek in F.C.B., Begoniaceae: 13 (1969); Kupicha in F.Z. 4: 500 (1978); Cribb & Leedal, Mountain Fl. S Tanzania: 55, t. 8a (1982); U.K.W.F. ed. 2: 92 (1994); Klazenga et al. in B.J.B.B. 63: 286, fig. 6, map (1994). Type: Angola, Golungo Alto, *Welwitsch* 875 (BM!, lecto.; COI, G, K!, iso.)

Erect, terrestrial herb 10–150 cm high (or scrambling up to 2 m), rarely branching; stem reddish or translucent, succulent, becoming woody with age; indumentum of simple hairs. Stipules elliptic to ovate, 1–3 cm long, caducous, boat-shaped, apex acuminate, with minute scattered glandular hairs. Leaves green above, reddish or purplish beneath, drying papery, asymmetrical, commonly 3–5-lobed to almost palmatifid, sometimes transversely broadly ovate with 5 apices, 7–28 × 10–19 cm, base obliquely cordate, margin denticulate to serrulate to almost entire, if leaves not lobed then with 5 acute to caudate points (large teeth), apex of lobes acute to acuminate, if leaves not lobed then apex caudate to acuminate, sparsely puberulent to glabrous; venation palmate; petiole reddish, 8–24 cm long. Inflorescence axillary, unisexual or bisexual, a dichasial cyme with up to 31 flowers; peduncle 1–4 cm long; each ramification subtended by a pair of caducous ovate to broadly ovate acuminate bracts 0.7–1.4 cm long, chartaceous, with conspicuous veins; pedicels 0.3–2.3 cm long; bracteoles absent. Male flowers: tepals 2(–4), white to pink with darker venation, perianth cylinder absent; outer tepals broadly ovate, 0.5–1.4 × 0.4–1.6 cm; inner, if present, narrowly elliptic, less than 2 mm long; stamens 7–35, filaments free or fused only at the base, anthers narrowly obovoid, 1.5–2.5 mm long. Female flowers: tepals 2(–4), the outer and inner very similar to those in male flowers; ovary ellipsoid, wings absent, 3-locular; placentation parietal through 60% of the ovary; styles 3, forked at the apex, spirally twisted once, glabrous; stigmatic band yellowish green. Fruits greenish tinged with red, fleshy, ellipsoid or trigonous, 2–3 cm long, 0.6–1.2 cm in diameter, terete to triangular in cross section with narrow longitudinal ribs or furrows, indehiscent, styles and tepals often persistent until complete maturity. Fig. 3, p. 16.

UGANDA. Bunyoro District: Budongo forest, Dec. 1933, *Eggeling* 1409!; Toro District: W side of Bwamba Pass, 27 Jan. 1935, *G. Taylor* 3272!; Kigezi District: Ruwenzori, Mihunga, 11 Jan. 1939, *Loveridge* 338!

KENYA. N Kavirondo District: Kakamega Forest, Kibiri block, 21 Jan. 1970, *Faden et al.* 70/40!

TANZANIA. Lushoto District: E Usambara, Kwamkoro Forest Reserve, 27 Oct. 1986, *Iversen et al.* 86/200!; Morogoro District: Uluguru Mts, Lupanga peak, 22 May 1933, *B.D. Burtt* 4720!; Iringa District: Mufindi, Kigogo R., 19 Mar. 1962, *Polhill & Paulo* 1819!

DISTR. **U** 2, 4; **K** 5; **T** 1–3, 6, 7; widespread in tropical Africa from Sierra Leone and Guinea throughout Central Africa to Mozambique and Madagascar

HAB. Moist forest, especially near water; 600–1900 m

CONSERVATION NOTES. Least concern (LC); widespread

SYN. *B. heddei* Warb. in Gartenflora 49: 1, fig. 1470 (1900). Type: Tanzania, Lushoto District: W Usambara, Wilhelmstal, *Hedde* s.n. (B, holo.)

B. kummeriae Gilg in E.J. 34: 87 (1904). Type: Tanzania, Lushoto District: E Usambara, Nguelo, *Engler* 647 (B, holo.)

B. petrophila Gilg in E.J. 34: 86 (1904). Type: Tanzania, Lushoto District: W Usambara, Sakare, *Busse* 349 (B, holo.)

B. seretii De Wild. in Ann. Mus. Congo, Bot., sér. 5, 2: 59, fig. 18 (1907). Type: Congo-Kinshasa, between Mambanbu and Gangara, near Arebi, *Seret* 1216 (BR, holo.)

B. pycnocaulis Irmsch. in E.J. 81: 180 (1961). Type: Morogoro District: Uluguru Mts, *Schlieben* 3006 (B, holo.)

NOTE. There is extensive leaf variation within this species ranging from almost entire to palmatifid. There is a regional tendency for southern specimens like those from Tanzania or Mozambique to possess leaves which are almost entire, while palmatifid or lobed leaves are found in Kenya and Uganda. However, this is a variable character in itself with most specimens intermediate.

3. **Begonia ampla** *Hook.f.* in Oliv., F.T.A. 2: 574 (1871); Wilczek in F.C.B., Begoniaceae: 8 (1969); de Wilde & Arends in Misc. Papers Landb. Wageningen 19: 385, fig. 1 (1980). Type: Equatorial Guinea, Fernando Po, *Mann* 314 (K!, lecto.; K!, iso.)

Epiphytic herb with stems to 2 m long, growing on trees or rocks; stems creeping, rooting at the nodes, not or rarely branched, up to 1.8 cm in diameter, becoming woody with age; indument of translucent sessile peltate-helicoid scales with fimbriate or long dentate margins or stellate hairs. Stipules narrowly ovate to triangular, 3–5 cm long, caducous, boat-shaped, not keeled, apex acute to mucronate, almost glabrous to densely indumented. Leaves thick and fleshy, obliquely ovate to obliquely suborbicular, (8–)14–30 × (6–)10–25 cm, base cordate, margin entire to broadly denticulate (with minute teeth extending from lateral veins), apex commonly shortly caudate to acuminate; venation palmate, glabrescent to sparsely indumented above, glabrescent to densely indumented beneath, particularly on venation; petiole 5–23 cm long, glabrous or evenly pubescent. Inflorescence axillary, bisexual, the axes reduced, many-flowered; peduncle (0.5–)1–4 cm long; bracts 2, persistent, subtending and enveloping the inflorescence, overlapping, boat-shaped, 1.5–2.8(–3.5) cm long; bracteoles absent. Male and female flowers with only the free portion of the tepals exserted from the bracts, the base of the tepals fused to form a perianth cylinder up to 16 mm long. Male flowers: tepals 2, pink or white, obovate to broadly obovate, 1.6–4.3 × 1.3–4 cm, often with longitudinal dark pink striations; stamens 25–60, filaments free or fused only at the base, anthers oblong, 2.5–4 mm long. Female flowers: tepals 2, pink or white, obovate to nearly orbicular, 1.7–3.3 × 1.4–3.5 cm; ovary shortly stipitate, obovoid, 4-locular, without wings; placentation pseudo-axillary; styles 4, caducous in fruit, forked at the apex, spirally twisted once, with yellow to greenish stigmatic papillae. Fruits developing within the bracts, mature fruits white to orange, fleshy, baccate, obovoid to globose, 1–1.3 cm long, apiculate at the apex, 4-lobed in cross section, indehiscent.

UGANDA. Kigezi District: Ishasha Gorge, Kayonza Forest, Oct. 1940, *Eggeling* 4196! & same loc., no date, *Paulo* 664! & same loc., Mar. 1946, *Purseglove* 2006!

DISTR. **U** 2; Cameroon, Gulf of Guinea Islands, Gabon, Congo-Brazzaville, Congo-Kinshasa, Rwanda

HAB. Moist forest, 1200–1800 m

CONSERVATION NOTES. Least concern (LC); widespread

SYN. *B. duruensis* De Wild. in Ann. Mus. Congo, sér. 5 (Bot.) 2: 318 (1908). Type: Congo-Kinshasa, Chief Mugdangba's territory, *Seret* 544; & near Nala, 5 Aug. 1906, *Seret* s.n.; & Sabona, 9 June 1906, *Seret* s.n. (all BR, syn.)

NOTE. One of two species in this region characterised by the conspicuous overlapping, boat-shaped bracts that envelope the inflorescence and the fruits (the other being *B. poculifera*). Seeds seen to be dispersed by ants in Cameroon (Plana, pers. obs.).

4. **Begonia poculifera** *Hook.f.* in Oliv., F.T.A. 2: 574 (1871); de Wilde & Arends in Misc. Papers Landb. Wageningen 19: 392, fig. 3 (1980). Type: Cameroon, Cameroon mountains, *Mann* s.n. (K!, lecto.; P, iso.)

Epiphyte, growing on trees, rarely on rocks; stems creeping, rooting at the nodes, usually branched, up to 2 m in length and 1.2 cm in diameter, becoming woody with age; indumentum of translucent sessile peltate-helicoid scales with a dentate margin, the margin frequently curling under in herbarium specimens and appearing entire; hairs on young organs (e.g. new leaves, bracts) almost stellate. Stipules up to 5.5 cm long, caducous, narrowly ovate to narrowly triangular, boat-shaped, keeled, apex acute to obtuse, almost glabrous to densely indumented. Leaves thick and fleshy, broadly falcate to obliquely fan-shaped, (3.5–)9–15(–22) × (1.8–)4–11(–19) cm, base cuneate, margin entire, apex acuminate, rarely broadly denticulate (with minute teeth extending from lateral veins), when young densely covered in minute translucent scales, glabrescent with age; venation reddish beneath, palmate-pinnate; petiole 1–5.5 cm long. Inflorescence axillary, bisexual, the axes reduced, many-flowered; peduncle (0.5–)1.2–5.3(–7) cm long; bracts 2, persistent, subtending and enveloping the inflorescence, boat-shaped, overlapping, 1.5–2.4(–3) cm long, chartaceous, membranous at the margin; bracteoles absent. Male and female flowers with only the free portion of the tepals exserted from the bracts, the base of the tepals fused to form a perianth cylinder up to 12 mm long. Male flowers: tepals 2, white with longitudinal pink striations, broadly elliptic to orbicular, (0.4–)1.2–3.2 × (0.3–)1–2.8 cm; stamens 10–45, filaments free or fused only at the base; anthers oblong, ± 2.5 mm long. Female flowers: tepals 2, white, similar to male flowers in dimension; ovary shortly stipitate, obovoid, without wings, 4-locular; placentation pseudo-axillary; styles 4, forked at the apex, spirally twisted once, covered all over with minute yellow stigmatic papillae. Fruits developing within the bracts, mature fruits greenish-white often tinged with dark pink on the upper part, fleshy, baccate, obovoid, 0.8–2.3 cm long, 4–5-grooved, apiculate, indehiscent.

UGANDA. Ankole District: Kasyoha-Kitomi Forest Reserve, near Nzozia R., 13 Nov. 1994, *Poulsen et al.* 707!

TANZANIA. Morogoro District: Nguru Mts, Mkobwe, 29 Mar. 1953, *Drummond & Hemsley* 1897!; Iringa District: Luhega Forest Reserve, 19 Feb. 1997, *Hørlyck & Jøker* TZ328! & Udzungwa Mountain National Park, Mt Luhombero, 26 Sep. 2000, *Luke et al.* 6677!

DISTR. **U** 2; **T** 6, 7; Nigeria, Cameroon, Bioko, Gabon, Congo-Kinshasa, Rwanda, Burundi, Angola

HAB. Moist forest; 1250–1900 m

CONSERVATION NOTES. Least concern (LC); widespread

SYN. *B. haullevilleana* De Wild. in Ann. Mus. Congo, sér. 5 (Bot.) 2: 320 (1908); Wilczek in F.C.B., Begoniaceae: 9 (1969). Type: Congo-Kinshasa, between Zobia and Buta, *Seret* 866 (BR, holo.)

B. adolfo-friderici Gilg in Z.A.E.: 574 (1913); Wilczek in F.C.B., Begoniaceae: 10, t. 1 (1969)

NOTE. Easily identified by the characteristic falcate or fan-shaped leaves. One of two species possessing large boat-shaped bracts enveloping the inflorescence (see *B. ampla*).

5. **Begonia eminii** *Warburg* in E. & P. Pf. 3 (6a): 141 (1894) & in P.O.A. C: 282 (1895); F.P.S. 1: 183 (1950); Wilczek in F.C.B., Begoniaceae: 27, t. 3 (1969); U.K.W.F. ed. 2: 92 (1994); de Wilde in Wageningen Univ. Papers 2001-2: 71 (2002). Type: Tanzania, Bukoba, *Stuhlmann* 1453 (B, lecto., chosen by de Wilde)

Terrestrial or epiphytic, erect to 120 cm high or climbing (fide Vollesen) and to 4 m long; stems succulent, green and/or red, up to 1.2 cm thick, branched, rooting at the nodes, woody at the base; indumentum of peltate scales with fimbriate margins. Stipules narrowly triangular, 2–4.5 cm long, caducous, neither boat-shaped nor curved, apex acuminate, indumentum present on outside, glabrous within. Leaves drying papery, green above, purple or with reddish veins beneath, almost symmetrical, narrowly ovate to ovate, 6.5–22 × 2.5–11 cm, base oblique, deeply cordate, margin entire to obscurely denticulate, apex acute to shortly acuminate, glabrous above, sparsely hairy beneath, denser on venation; venation pinnate; petiole red, 2–13(–17) cm long, densely hairy. Inflorescence axillary, a unisexual dichasial cyme. Male inflorescence with up to 15 flowers; peduncle 5–15 mm long; bracts opposite, up to 7 mm long, smaller in the upper parts of the inflorescence, chartaceous, almost connate; pedicels up to 4–11 mm. Male flowers: tepals 4, free, white often tinged with red at the apex or pinkish, the outer elliptic to obovate, 4–12 × 3–6 mm, the inner narrowly obovate, 3–9 × 1.5–3 mm; stamens 7–15, free or fused at the base, anthers oblong, curved inward, 1–2 mm long, apex obtuse. Female inflorescence with 2–3(–7) flowers, peduncle 1–8 mm long; bracts similar to those in male inflorescence; bracteoles present, ovate, 2–4 mm. Female flowers: sessile, tepals 4, free, shape and colour similar to male flowers, the outer 5–10.5 × 2.5–5 mm, the inner 3–8 × 1.5–2.5 mm; ovary red, fusiform, without wings, 3-locular; placentation septal; styles 3, yellow, forked at the apex, glabrous, not persistent in fruit; stigmatic band yellow, horseshoe-shaped, weakly twisted ($^1/_2$ turn). Fruits pendulous when mature, red, fleshy, fusiform, 2–5 cm long, terete in cross section, straight, dehiscent by 1 or more lateral longitudinal slits.

UGANDA. Toro District: Kibale National Park, near Kanyawara, 15 July 1994, *Poulsen et al.* 683!; Mengo District: Kasa forest, 11 km S of Mitanya, 4 Oct. 1949, *Dawkins* 418! & Busiro, 1 km E of Jungo church, 30 Apr. 1969, *Lye & Morrison* 2711!

KENYA. N Kavirondo District: Kakamega Forest, Aug. 1934, *Dale* 3250!

TANZANIA. Bukoba District: Minziro Forest Reserve, 4 July 2000, *Bidgood et al.* 4832! & same loc., Kinwa Kyaishemweru area, 14 July 2001, *Festo et al.* 1619!

DISTR. **U** 2, 4; **K** 5; **T** 1; Nigeria, Cameroon, Central African Republic, Bioko, Equatorial Guinea, Gabon, Congo-Brazzaville, Congo-Kinshasa, Burundi, Sudan, Angola

HAB. Terrestrial or on living trees, logs or rocks in moist forest, often near streams, also in water-logged forest; 1000–1500 m

CONSERVATION NOTES. Least concern (LC); widespread

SYN. *B. poggei* Warb. in E.&P. Pf. 3(6a): 141, fig. 48h (1894) & in E.J. 22: 35 (1895). Type: Congo-Kinshasa, Mukenge, Tschingari-gari, *Pogge* 962 (B, lecto.)

B. poggei Warb. var. *albiflora* Th. & Hel. Durand, Ann. Mus. Congo 1: 104 (1908) & Syll. Fl. Congol.: 234 (1909). Type: Congo-Kinshasa, Mutumbi, Ponthier-ville-Kindu, *Dewevre* 1092 (BR, holo.)

B. ealensis Irmsch. in E.J. 57(2): 241 (1921). Type: Congo-Kinshasa, botanical gardens at Eala, *Chevalier* 28046 (P, holo.)

6. **Begonia kisuluana** *Büttner* in Verhandl. Bot. Ver. Prov. Brandenburg 32: 45 (1890); Wilczek in F.C.B., Begoniaceae: 22 (1969); de Wilde in Wageningen Univ. Papers 2001-2: 124, fig. 12, map (2002). Type: Angola, Arthington Falls near Kisulu, three days journey from San Salvador, *Büttner* 496 (B, lecto.)

Epiphytic herb or soft-wooded shrub with stems to 1(–2) m long; stems erect, up to 6 mm thick, branched, rooting at the nodes, woody at the base; indumentum of peltate scales with fimbriate margins. Stipules narrowly triangular, 1–2.5 cm long, caducous, boat-shaped, not curved, apex acuminate, hairy on outside. Leaves leathery, drying thin, weakly asymmetrical, ovate to elliptic, sometimes narrowly elliptic, 2–12 × 0.8–6.2 cm, base obliquely cordate to rounded or truncate, margin entire, rarely obscurely denticulate towards the apex, apex commonly broadly acute to obtuse, rarely acuminate, glabrescent above and below, with a few scattered lepidote hairs; venation pinnate; petiole 0.4–2.2(–6) cm long. Inflorescence axillary, a unisexual dichasial cyme. Male inflorescence with up to 15 flowers, peduncle (0.4–)0.8–2.8(–3.5) cm long; each ramification subtended by a pair of opposite bracts 3–7 mm long, smaller in the upper parts of the inflorescence, chartaceous, almost connate; pedicels 0.4–1.2 cm long; bracteoles absent. Male flowers: tepals 4, free, white suffused with pink on the upper half, the outer obovate, 5–8 × 3–5.5 mm, the inner narrowly obovate, 3.5–8 x 1–3 mm; stamens 10–19, filaments fused 0.8–1.5 mm from the base; anthers narrowly ellipsoid to obovoid, 0.8–1.5 mm long, apex obtuse to acute. Female inflorescence with 1–3 flowers, peduncle 1–5 mm long; bracts similar to those in male inflorescence but up to 8 mm long; pedicel to 2 mm long. Female flowers: subsessile, tepals 4, free, shape and colour similar to male flowers, the outer 6–12 × 3–5.5 mm, the inner 4–10 × 1.5–3 mm; ovary fusiform, without wings, 3-locular; placentation parietal through less than 40% of the ovary, the rest septal; styles 3, simple at the apex (1 flower observed to have forked style), glabrous, not persistent in fruit; stigma yellow, simple. Fruits fleshy, fusiform, often somewhat curved, 1.2–3.5 cm long, 0.3–0.4 cm in diameter, terete in cross section, greenish-yellow, dotted with lenticels, dehiscent by a lateral longitudinal slit.

UGANDA. Bunyoro District: Budongo Forest, May 1935, *Eggeling* 2013! & 2082! & same loc., compt. 41, 15 Sep. 1977, *Katende* K2765!

DISTR. **U** 2; Nigeria, Cameroon, Gabon, Congo-Brazzaville, Congo-Kinshasa, Angola

HAB. Swampy river valleys within forest; altitude not given, presumably around 700–1300 m

CONSERVATION NOTES. Least concern (LC); widespread

SYN. *B. zobiaensis* De Wild. in Ann. Mus. Congo, Bot., Ser. 5(2): 324 (1908). Type: Congo-Kinshasa, between Zobia and Buta, *Seret* 882 (BR, holo., iso.)

NOTE. Although the species is described as possessing simple, unforked styles, a single flower from *Katende* K2765 has forked styles with a long horseshoe-shaped stigma, almost identical to those in *B. eminii*. This however, appears to be an aberrant flower; the other flower on the same inflorescence has typical *B. kisuluana* styles.

7. **Begonia tatoniana** *Wilczek* in B.J.B.B. 39: 89, fig. 1 (1969) & in F.C.B., Begoniaceae: 24, fig. 2 (1969); de Wilde in Wageningen Univ. Papers 2001-2: 236, fig. 29, map 29 (2002). Type: Congo-Kinshasa, Kasai, Kiyaka, Kwango, Lutete forest (BR, holo.; K!, iso.)

Epiphyte, semi-erect or trailing; stems to 80 cm long, up to 7 mm thick, branched, rooting at the nodes, woody at the base. Indumentum of peltate scales with fimbriate margins. Stipules narrowly triangular to triangular, 0.8–1.5 cm long, caducous, apex acuminate, indumentum present on outside. Leaves drying papery, ± symmetrical, narrowly elliptic to narrowly ovate, 3.5–15 × 1.3–5 cm, base obliquely and weakly cordate, margin entire to remotely denticulate, apex acute to shortly acuminate, glabrous above, sparsely scaly beneath, dense on venation; venation pinnate; petiole 0.8–3.5 cm long, puberulous. Inflorescence axillary, a bisexual dichasial cyme, 3–flowered with 2 lateral ♀ and a central ♂; peduncle 2–20 mm long, densely scaly; each ramification subtended by a pair of bracts; bracts fused into a chartaceous caducous bifid ochrea, 3–7 mm long, splitting with age; small opposite bracts supporting the female flowers, looking like bracteoles; pedicels (♂ only) 2–14 mm long; bracteoles absent. Male flowers: tepals 4, free, white or pinkish, the outer ovate, 7 × 3 mm, the inner narrowly ovate, 3–4.5 × 1.5–2 mm; stamens 9–20, filaments fused at the base to form a column, anthers oblong, 1–1.5 mm long. Female flowers: sessile, tepals 4, free, colour and shape as in the male flowers, the outer 0.6–1 × 0.3–0.6 cm long, the inner 0.4–0.7 × 0.15–0.4 cm; ovary fusiform, 8–13 mm long, without wings, 3-locular; placentation unknown, probably septal; styles 3, forked at the apex, glabrous, not persistent in fruit; stigmatic band yellow, horseshoe-shaped, weakly twisted. Mature fruits sessile (2 per infructescence), red, fleshy, fusiform, 2.2–3.9 cm long, 0.2–0.4 cm in diameter, terete in cross section, dehiscence unknown but probably by one or more longitudinal slits, with a distinct apical scar where tepals and styles have fallen off.

UGANDA. Masaka/Mengo District: near mouth of Kagera R., 11 Feb. 1904, *Bagshawe* 564!; Masaka District: Lake shore, Misozi, [Musozi], 11 Dec. unknown year, *Bagshawe* 2!
DISTR. **U** 4; Congo-Kinshasa, Angola
HAB. Presumably moist forest; 1100–1150 m
CONSERVATION NOTES. Probably least concern (LC)

8. **Begonia zimmermannii** *Irmsch.* in E.J. 81: 181 (1961); de Wilde in Wageningen Univ. Papers 2001-2: 242, fig. 30, map 30 (2002). Type: Tanzania, Lushoto District: Usambara, Amani, *Greenway* 1045 (K!, holo.; B, EA!, K!, PRE, WAG, iso.)

Epiphytic herb, erect or reclinate; stems up to 1.5 cm thick, little branched, rooting at the nodes in older stems; indumentum of peltate scales with fimbriate margins. Stipules narrowly triangular ± 3.5 cm long, caducous, weakly boat-shaped, curved, apex acuminate, indumentum present on outside, glabrous within. Leaves thick and fleshy, leathery, asymmetrical, weakly falcate, obliquely ovate to elliptic, 10–15 × 5–8 cm, base weakly cordate to rounded, margin entire, apex obtuse to shortly acuminate, glabrescent above and below, with a few scattered hairs; venation pinnate; petiole 1.5–2.8(–4) cm long. Inflorescence axillary, a bisexual dichasial cyme, with up to 11 flowers; peduncle 0.5–1.7 cm long, densely squamose-stellate hairy; each ramification subtended by a pair of opposite weakly boat-shaped elliptic bracts 2–2.5 cm long, chartaceous, with conspicuous veins, apex acuminate, caducous; pedicels (♂ only) 0.5–1.2 cm long; bracteoles absent. Male flowers: tepals 4, free, the outer broadly elliptic, 0.7–1.3 × 0.6–1 cm, the inner obovate, boat-shaped, 0.5–0.7 × 0.3–0.4 cm; stamens 17–19, filaments free or fused at the base; anthers oblong, 1–2 mm long. Female flowers: tepals 4, free, white, the outer orbicular, 0.4–0.7 cm in diameter, the inner 0.2–0.35 × 0.15–0.2 cm; ovary fusiform, without wings, 4-locular; placentation unknown, probably parietal; styles 4, shortly forked at the apex, glabrous, not persistent in fruit; stigmatic band yellow, horseshoe-shaped, not twisted. Fruits sessile, fleshy, fusiform, 4.6–5(–6) cm long, terete in cross section, indehiscent, with a distinct apical scar where tepals and styles have fallen off.

TANZANIA. Lushoto District: Usambara, Amani, 18 Dec. 1928, *Greenway* 1045! & idem, *Grote* 3694 & Tea estate NE of Amani, 5 Oct. 1979, *Kibuwa* 5194!
DISTR. **T** 3; not known elsewhere
HAB. Moist forest; ± 900 m
CONSERVATION NOTES. Vulnerable (VU); very rare in its single locality, known only from three specimens

9. **Begonia engleri** *Gilg* in E.J. 34: 97 (1905); Irmsch. in E.J. 81: 147 (1961). Type: Tanzania, Lushoto District: E Usambara, Sigi, *Engler* 640 & 841 (B, syn.)

Herb to 120 cm, usually epilithic, prostrate basally; stem red- or purple-flecked, fleshy, thickest at base, not branched, 13–15 mm in diameter, with many long purple hairs or glabrous. Stipules persistent, broadly ovate, 11–15 mm long, acute, glabrous. Leaves red- or purple-flecked, slightly fleshy, oblong-ovate, oblique, 9–30 × 4.5–13 cm, one side 1.5–2 times wider than the other, the larger with a basal rounded lobe with 3–4 main veins $^1/_3$ to $^1/_2$ of the lamina length, base cordate, margin deeply doubly-serrate, apex shortly acuminate, sparsely pilose; venation palmate-pinnate, the main vein at an angle to the petiole; petiole 3–14 cm long, with rings of hairs or fimbriate trichomes to 8 mm long near junction with lamina. Inflorescence an erect axillary dichasial cyme; peduncle purple-flecked, 5–17 cm long, glabrous or sparsely pilose; bracts soon caducous, membranous, broadly ovate, 7–9 mm long; pedicels of ♂ 1–2.7 cm long, of ♀ to 1 cm long; female flowers with two small bracteoles at base. Male flowers: tepals 4, pink, the outer obovate and 9–14 × 6–14 mm, the inner slightly smaller; stamens many, filaments slightly connate at base, anthers obovoid. Female flowers: tepals 5, pink, obovate-orbicular, 5–9.5 × 3–6 mm; ovary ellipsoid, 4–7 mm long, 3-locular, 3-winged; placentation axillary; styles 3, divided halfway, the branches twisted, persistent in fruit; stigmatic band continuous, spiral. Fruit a 3-winged brown papery capsule 1.2–2.6 cm long, the wings usually sub-equal, triangular, (0.6–)0.7–2.4 cm wide, rounded, dehiscing at the juction with the wings.

TANZANIA. Lushoto District: E Usambara, Kwamkoro–Kihuhuwi track, 16 July 1940, *Greenway* 5970! & Kisiwani, 30 May 1950, *Verdcourt* 250! & Muheza to Amani, 16 Apr. 1968, *Renvoize & Abdallah* 1525!

DISTR. **T** 3; not known elsewhere

HAB. Moist rock faces in the forest zone (or once in the dry bushland zone); (350–)800–1800 m

CONSERVATION NOTES. Vulnerable (VU)

SYN. *B. engleri* Gilg var. *nuda* Irmsch. in E.J. 81: 149 (1961). Type: Tanzania, Lushoto District: Longuza–Sigi road, *Greenway* 722 (K!, holo.; EA!, iso.)

10. **Begonia johnstonii** *Hook.f.* in Bot. Mag. 112: t. 6899 (1886); Irmsch. in E.J. 81: 149 (1961); Cribb & Leedal, Mountain Fl. S Tanzania: 53, pl. 7a (1982); U.K.W.F. ed. 2: 92 (1994). Type: see note.

Herb, possibly perennial, erect, procumbent or straggling, 20–60 cm high or to 90 cm long; stem swollen at base, succulent higher up, green with red streaks, glabrous. Stipules persistent, reddish, ovate to broadly ovate, 3–8 mm long, with entire margins. Leaves green above (and sometimes glaucous), green or often reddish beneath, ovate in outline, asymmetric, 5–18 × 2–8.5 cm, one side larger, up to 1.3 times the width of the other and with a rounded basal lobe with 2–3 main veins and $^1/_4$ to $^1/_3$ the total length of the lamina, base cordate, margin crenate, apex acuminate, sparsely pilose or glabrous; venation palmate-pinnate, the main vein at an angle to the petiole; petiole often reddish, or with red spots, (3–)6–16 cm long, with ring of white hairs at the junction with the lamina. Inflorescence an axillary dichasial cyme, with up to 6(–12) flowers; peduncle reddish, (3–)6–23 cm long, glabrous; each ramification subtended by a pair of opposite bracts 2–5 mm long, membranous; pedicels 5–23 long with 2 minute bracteoles on those of the ♀ flowers. Male flowers: tepals 4, free, pink to white, the outer obovate, 6–15 × 4–10 mm, the inner obovate, 7–19 × 5–13 mm; stamens numerous, filaments free, 1–4 mm long, anthers ellipsoid to subglobose, 0.5–1 mm. Female flowers: tepals 5, pink to white, obovate, 5–13 × 3–8 mm; ovary globose to ellipsoid, 7–8 mm long, 3-locular, 3-winged; placentation axile; styles 3, shortly forked halfway or at apex, glabrous, persistent; stigmatic band yellow, horseshoe-shaped, twisted. Fruit a brown capsule with red wings, globose, ovoid or ellipsoid, 7–11 × 6–9 mm excluding the wings; wings very unequal, one laterally triangular and 12–27 mm wide, the other two triangular, rounded, 6–8 mm wide; dehiscence along the junction with the wings. Fig. 2, p. 5.

KENYA. Northern Frontier District: Mt Nyiru, Apr. 1995, *Bytebier et al.* 324!; Meru District: Karita River near Boma, Sept. 1951, *Archer* in *Bally* B8038!; Teita District: Kasigau, Bungule route, 18 Nov. 1994, *Luke* 4181!

TANZANIA. Lushoto District: W Shagai forest, 24 May 1953, *Drummond & Hemsley* 2728!; Morogoro District: Lupanga Peak, Aug. 1951, *Greenway* 8606!; Iringa District: N part of Gologolo Mts, 13 Sept. 1970, *Thulin & Mhoro* 957!

DISTR. **K** 1, 4, 6, 7; **T** 2, 3, 6, 7; not known elsewhere

HAB. On steep rocks in forest or in moist sites, less often in forest without any specification; local, rarely locally common; 750–2400 m

CONSERVATION NOTES. Least concern (LC)

SYN. *B. johnstonii* Hook.f. forma *pilosa* Irmsch. in E.J. 81: 151 (1961). Type: Tanzania, Uluguru Mts, *Schlieben* 2807 and 2927 (B, syn.), **syn. nov.**

NOTE. Typification unclear: the protologue starts with an 'ined.' Latin description of *B. johnstonii* by Oliver, based on *Johnston* s.n. from Kilimanjaro, '5000–6000 feet'; the plate and main description (in English) by Hooker are based on plants grown at Kew, flowered in April 1886, from material sent by Bishop Hannington from Masailand – and there is a specimen dated 1883 at Kew. Irmscher states the Johnston plant is the holotype. We incline towards the Kew ones.

11. **Begonia schliebenii** *Irmsch.* in E.J. 81: 146 (1961). Type: Tanzania, Morogoro District: Uluguru Mts, Mbambaku, *Schlieben* 3584 (B, holo.; K!, iso.)

Herb to 13 cm high; stems creeping then ascending, 5–8 mm in diameter, densely leafy, covered in stipules, glabrous. Stipules persistent, membranous, broadly reniform, 12–30 × 23–50 mm, margin entire. Leaves membranous, asymmetrical, obovate-oblong to oblong, 8–17 × 3.5–9 cm, base obliquely cordate, margin crenate-serrate, apex acuminate, pilose and often purplish above and beneath; venation pinnate; petiole 4–11 cm long, sparsely hairy and with a ring of hairs 3–8 mm long at the junction with the lamina. Flowers few in long cymes; peduncle 6–19 cm long; bracts broadly ovate, 7 × 10–16 mm, persistent; pedicels 20–25 mm long, with very small bracteoles on those of the ♀ flowers. Male flowers: tepals 4, (colour not known), the outer ovate, 14 × 9 mm, the inner 12 × 7 mm; stamens many, filaments 1–2 mm long, free, anthers ± 0.8 mm long. Female flowers: tepals 5, (colour not known), ovate, 10 × 7 mm; ovary ovoid, 8 × 8 mm, 3-locular, 3-winged; placentation axile; styles three, branched halfway, the branches twisted; stigmatic band continuous, spiral. Fruit an ovoid capsule, 9 × 9 mm excluding the wings, one wing to 23 mm long, the other two smaller; dehiscence along the junction with the wings.

TANZANIA. Morogoro District: Uluguru Mts, Mkambaku NE, 26 Feb. 1933, *Schlieben* 3584!
DISTR. **T** 6, 7; not known elsewhere
HAB. Moist rock face in forest zone; ± 2040 m
CONSERVATION NOTES. Data deficient (DD) but at least Vulnerable (VU)
NOTE. Recently collected by *Luke et al.* 6588, 7832 and probably 11342 (large stipules, but stem covered in red hairs); and *Festo et al.* 2120 (fide Luke). Additional measurements from these specimens: plant to 40 cm high; stipules papery; petiole with pink apical hairs.

12. **Begonia wollastonii** *Baker f.* in J.L.S. 38: 252 (1908), as *wollastoni*; Irmsch. in E.J. 81: 152 (1961); Wilczek in F.C.B., Begoniaceae: 47, t. 5 (1969); Sands et al. in Fl. Eth. 2, 2: 60 (1995). Type: Uganda, Ruwenzori, 2100 m, March 1906, *Wollaston* s.n. (BM!, holo.)

Fleshy herb up to 1.2(–2.4) m high, terrestrial or occasionally on rock; stem erect, juicy, green, to 2 cm across, glabrous, with short branches and often with axillary bulbils; rootstock usually tuberous, red, to 7.5 cm in diameter. Stipules ovate to elliptic, 4–23 mm long, with entire margins. Leaves light green, ovate in outline, very asymmetric, 5–20 × 3–16 cm, base cordate, margins dentate-serrate, apex slightly acuminate, sparsely hairy with white or red hairs on both surfaces; venation palmate; petiole 2–20 cm long, with ring of hairs/fimbriate trichomes at junction with lamina. Flowers in axillary unisexual or bisexual cymes with 1–2 dichasia; peduncle 10–55 mm long; bracts caducous, ovate, 7–12 × 3–6 mm; pedicels 10–30 mm long; bracteoles absent. Male flowers: tepals 4, free, pink or white, the outer broadly ovate, suborbicular or ovate, 17–33 × 17–30 mm, the inner elliptic to obovate, 10–22 × (5–)10–14 mm; stamens many (> 40), filaments 1–5 mm long, anthers ellipsoid, 0.8–1.4 mm. Female flowers: tepals (4–)5, pink or white, the outer four elliptic to oblong, 10–17(–27) × 7–15 mm, the innermost smaller; ovary ovoid, 6–12 × 2–8 mm, 3-locular, 3-winged; placentation axile; styles three, to 6 mm long, divided halfway, the branches twisted; stigmatic band continuous, spiral. Fruit an ovoid or ellipsoid capsule, 15–28 × 9–20 mm excluding the wings; wings very unequal, the largest triangular and 15–40 mm wide, the other two much smaller; dehiscence usually along the junction with the wings.

UGANDA. Karamoja District: Mt Kadam above Moruito, without date, *Symes* 634!; Mt Elgon, Bulago, Bugishu, 9 Dec. 1938, *A.S. Thomas* 2578! & Sasa trail, 27 Dec. 1996, *Wesche* 600!
KENYA. Mt Elgon East, Dec. 1957, *Tweedie* 1482!; Mt Kenya, Katito Forest, Karinga R., June 1939, *Mrs H. Copley* in *Bally* 100! & Kamweti track above Castle Forest Station, 31 Jan. 1971, *Faden et al.* 71/106!
TANZANIA. Mbeya District: Mbeya Range, Dec. 1979, *Leedal* 5813! & Kitulo Plateau, Ishinga Mt, Feb. 1979, *Cribb et al.* 11357!

DISTR. **U** 1–3; **K** 3, 4; **T** 7; Congo-Kinshasa, Ethiopia
HAB. In moist forest along streams and in spray zone of waterfalls, always in very wet sites where it may be locally common; (1350–)1750–2900 m
CONSERVATION NOTES. Least concern (LC); widespread in a common habitat

SYN. *B. abyssinica* Cufod. in Senck. Biol. 41: 382 (1960). Type: Ethiopia, Mt Dida, *Kuls* 718 (FR, holo.)
B. keniensis Engl. in Veg. Erde 9, 3, 2: 621 (1921); Irmsch. in E.J. 81: 153 (1961); U.K.W.F. ed. 2: 92 (1994). Type: none mentioned, "Knollenpflanze, am Kenia um 2300 m", possibly *Battiscombe* 551 or 136 (see note)
B. lebrunii Robyns & Lawalrée in B.J.B.B. 18: 285 (1947). Type: Congo-Kinshasa, Ruwenzori, Mt Wandundu, *Lebrun* 4434 (BR, holo.)

NOTE. Hitherto *B. wollastonii* and *B. keniensis* have been kept apart on flower colour (but within single populations both colours can occur: *A.S. Thomas* 2578, Mt Elgon), tepal shape in male flowers longer than broad, or broader than long (but variation is continuous, with suborbicular intermediates!) and the size of the largest fruit wing - long in the 'type' of *keniensis* (but again, intermediates occur). Based on these observations, we have decided the two are synonymous; *B. wollastoni* is the older name.

Battiscombe 136 at Kew (from 'British East Africa', no further locality) bears a hand-written det-slip: "Begonia keniensis Gilg !1912, det. E.Gilg" and could be the original sheet from which Gilg coined the name – later taken up by Engler. *Battiscombe* 551, collected in 1912, bears label data stating it is from Mt Kenya, and that the rootstock is tuberous (but still without altitude). Engler cites the name as "*B. keniensis* Gilg" but as he (Engler) wrote the text the authorship is Gilg ex Engler; 'ex' authors are not cited in this Flora.

The two specimens from **T** 7 are curiously remote from the main population.

13. **Begonia princeae** *Gilg* in E.J. 30: 361, fig. (1902); Irmsch. in E.J. 81: 114 (1961); Wilczek in F.C.B., Begoniaceae: 44 (1969); Kupicha in F.Z. 4: 501 (1978). Type: Tanzania, Uhehe Plateau, Ubena, *Prince* 2 (B, holo.)

Perennial herb 5–60 cm high, erect and leafy, or, more rarely, scapose with basal leaves; stem simple or slightly branched, fleshy, red or pink, glabrous; basal tuber small, ellipsoid or globose, pink, covered in brown hairs. Stipules ovate, 5–22 mm long, with entire or remotely ciliate-serrate margins. Leaves green above, often red, purple or pink beneath, asymmetrically elliptic to suborbicular, 2–7.4 × 2–6(–8.4) cm, base truncate to slightly cordate, margin shallowly lobed, irregularly crenulate, apex acute to acuminate, glabrous; venation palmate; petiole 0.4–3.5(–5) cm long. Flowers in axillary and terminal bisexual cymes with 1–2 dichasia; peduncle 3–45 mm long; bracts ovate, 1–14 mm long, acute, persistent; pedicels 8–25 mm long; bracteoles absent. Male flowers: tepals 4, pink, the outer suborbicular, 9–16 × 8–14 mm, the inner obovate, 5–11 × 3–6 mm; stamens 40–60, filaments 1–3 mm long, free, anthers ellipsoid, ± 1 mm long. Female flowers: tepals (3–)5, pink, the outer four elliptic to oblong, 6–13 × 4–10 mm, the innermost smaller; ovary ovoid, 4–12 × 3–7 mm, 3-locular, 3-winged; placentation axile, bilamellate; styles three, 2.2–3.5 mm long, persistent, divided halfway, the branches twisted with a continuous spiral stigmatic band. Fruit pinkish, globose, ovoid or ellipsoid, 6–14 × 5–10 mm excluding the wings; wings very unequal, the largest broadly triangular, 11–15 mm wide, rounded to acute, the other two to 10 mm wide, rounded; dehiscence along the junction with the wing.

TANZANIA. Buha District: Gombe National Park [Kasakela Reserve], 17 Nov. 1962, *Verdcourt* 3342!; Iringa District: 40 km on Mafinga–Madibira road, 27 Jan. 1991, *Bidgood, Congdon & Vollesen* 1290!; Mbeya District: base of Pungaluma Hills between Songwe and Pungaluma plantations, 17 Dec. 1989, *Lovett, Sidwell & Kayombo* 3747!
DISTR. **T** 4, 7; Congo-Kinshasa, Angola, Zambia, Malawi, Mozambique
HAB. On steep rock or earth banks in moist and shaded places, occasionally on termitaria in woodland, twice described as an epiphyte; 750–1800(–2400) m
CONSERVATION NOTES. Least concern (LC); widespread

SYN. *B. verdickii* De Wild., Ann. Mus. Congo, Bot., 4: 93 (1903). Type: Congo-Kinshasa, Lukafu, *Verdick* 274 (BR, holo.)

B. homblei De Wild. in B.J.B.B. 5: 51 (1915). Type: Congo-Kinshasa, Lubumbashi [Elisabethville], *Homblé* 239; & Shinsenda, *Ringoet coll. Homblé* 400 (both BR, syn.)
B. subacuto-alata De Wild. in B.J.B.B. 5: 52 (1915). Type: Congo-Kinshasa, Lualaba kraal, *Homblé* 956 (BR, holo.)
B. princeae Gilg var. *princeae* forma *grossidentata* Irmsch. in E.J. 81: 119 (1961). Type: Zambia, above Mukoma to Inona Falls, *Richards* 3701 (K!, holo.)
B. princeae Gilg var. *princeae* forma *vulgata* Irmsch. in E.J. 81: 120 (1961). Type: Tanzania, Mbeya District: Utengule, *Stolz* 166 (B, holo.)
B. princeae Gilg var. *rhodesiana* Irmsch. in E.J. 81: 120 (1961). Type: Zambia, Mbala [Abercorn], Kambole Escarpment, *Richards* 8279 (K!, holo.)
B. princeae Gilg var. *rhodesiana* Irmsch. forma *rhodesiana* Irmsch. in E.J. 81: 121 (1961)
B. princeae Gilg var. *rhodesiana* Irmsch. forma *racemigera* Irmsch. in E.J. 81: 122 (1961). Type: Zambia, Mbala [Abercorn], Kambole Escarpment, *Richards* 8255 (K!, holo.)
B. princeae Gilg var. *racemigera* (Irmsch.) Wilczek in F.C.B.: 45 (1969)

NOTE. As the synonymy suggests, there is considerable variation in habit and morphological characters. Some specimens are leafy throughout, others are almost scapose with two leaves appressed to the ground.

14. **Begonia riparia** *Irmsch.* in E.J. 81: 126 (1961). Type: Tanzania, Songea District: Mahanja [Mahenge], Mbangalaca, Mbanga R., *Schlieben* 1806 (B, holo.)

Herb 15–25 cm high; stem erect or curved upwards from base, scarcely branched, with few leaves. Stipules persistent, membranous, oblong or oblong- lanceolate, 3–8 mm long, margin irregularly glanduliferous, apex obtuse or acute. Leaves asymmetric, broadly ovate to suborbicular, 4–7.5 × 4–8 cm, base obliquely cordate, margin crenate or crenulate, apex obtuse, glabrous or with a few hairs; venation palmate; petiole 3–10 cm long, glabrous. Flowers in axillary bisexual racemes with 3–4 lateral cymes; peduncle 10–18 mm long; bracts ovate to oblong, 1–3 mm long (?caducous/persistent?); pedicels 8–9 mm. Male flowers: tepals 4, white or pink, the outer oblong, 6–7 × 4–4.5 mm, the inner obovate, 5.5–5.6 × 2–2.5 mm; stamens number not known, staminal column 1.5–1.8 mm long, filaments 0.6–0.7 mm long, anthers obtrapeziform, 0.7 mm long. Female flowers: tepals 5, white or pink, oblong, 4.5–5.5 × 2–2.8 mm; ovary ovoid, 3 × 2 mm, 3-locular, 3-winged; placentation axile; styles three, not branched; stigmatic band lunate. Fruit a broadly ovoid capsule, 6 × 4 mm, wings ± equal, triangular, 3.5–5 mm wide, probably dehiscing along the junction with the wing.

TANZANIA. Songea District: Mahanja [Mahenge], Mbangalaca, Mbanga R., Feb. 1932, *Schlieben* 1806 (type)
DISTR. **T** 7, 8; not known elsewhere
HAB. Under riverine trees; ± 600 m
CONSERVATION NOTES. Data Deficient (DD)

NOTE. Recently collected from Udzungwa Mountain National Park, Matundu forest (**T** 7) by *Luke et al.* 11504; largest leaves 9.5 × 9.5 cm; some leaves mottled with white.

15. **Begonia stolzii** *Irmsch.* in E.J. 81: 174 (1961). Type: Tanzania, Kyimbila, falls by Mwakele, *Stolz* 685 (B, holo.; K, iso., not found)

Herb 15–50 cm high; stem erect or curved from the base, glabrous. Stipules persistent, membranous, ovate, 2.5–4 mm long, slightly fimbriate at the margin, acute. Leaves membranous, ovate, 4.5–7 × 2–3.5 cm, base obliquely cordate, apex acuminate, margin serrate, pilose above, almost glabrous beneath; venation ± palmate; axillary bulbils may be present; petiole to 5.5 cm long, with ring of hairs at the junction with the lamina. Flowers in axillary bisexual cymes, 2–4-flowered; peduncle 1.5–2.8 cm; bracts ovate, 3–5 mm long, persistent; pedicels 10–30 mm long. Male flowers: tepals 4, pink, the outer orbicular or ovate, 9 × 6–9 mm, the inner 7–9 × 4–6.5 mm; filaments 1–1.5 mm, anthers ellipsoid, 1.2–1.5 mm long. Female flowers: tepals 5, (colour not known), oblong, 6–8 × 2.8–5 mm; ovary ovoid, 7.5 × 4 mm, 3-locular, 3-winged;

FIG. 3. *BEGONIA OXYLOBA* — **1**, habit. *BEGONIA SUTHERLANDII* — **2**, habit. From Cribb & Leedal, Mountain Fl. S Tanzania. Drawn by Mair Swann, and reproduced with permission.

placentation axile, placentas undivided; styles three, caducous, branched halfway, the arms twisted, with a continuous, spiral stigmatic band. Fruit an ovoid to subglobose capsule, 4–9 × 3.5–4 mm, the wings subequal and to 11 mm wide; dehiscence along the junction with the wing.

TANZANIA. Njombe District: Lupembe, upper Ruhudji R., Feb. 1931, *Schlieben* 135!; Rungwe District: Kyimbila, falls by Mwakele, *Stolz* 685

DISTR. **T** 7; not known elsewhere

HAB. Streamside; ± 1000 m

CONSERVATION NOTES. Data Deficient (DD)

NOTE. Irmscher states that this is very close to (*B. bequaertii* and) *B. sutherlandii*, but the characters he mentions seem to be restricted to differences in leaf margin dissection. In due course, *B. stolzii* may be better regarded as part of the range of variation of *B. sutherlandii*.

16. **Begonia sutherlandii** *Hook.f.* in Bot. Mag. 94: t. 5689 (1868); Irmsch. in E.J. 81: 162 (1961); Wilczek in F.C.B., Begoniaceae: 50 (1969); Kupicha in F.Z. 4: 503 (1978); Cribb & Leedal, Mountain Fl. S Tanzania: 55, pl. 8b (1982). Type: specimen cultivated in UK, York, June 1867, *Backhouse* s.n. (not found)

Perennial herb 10–50 cm high, epiphytic or terrestrial; tuber depressed-ellipsoid, to 6 cm in diameter, red or orange inside; stem often red, sparsely branched, fleshy, glabrous or sparsely hairy. Stipules pink, asymmetrically ovate-oblong, 3–15 mm long, irregularly ciliate-dentate or fimbriate, acute to acuminate. Leaves green above, the margin and veins red, reddish beneath, ovate to lanceolate in outline, asymmetric, 2.5–20 cm long, 1.5–14.5 cm wide, one side 1.5–2 times wider than the other and with a basal lobe often $^1/_4$ to $^1/_6$ of the total lamina length, base cordate and very unequal, margin often shallowly lobed, always serrate, apex attenuate, sparsely to densely hairy; venation palmate-pinnate; bulbils sometimes present in leaf-axils; petiole red, 1–10 cm long, with or without a ring of hairs at the junction with the lamina. Flowers in axillary bisexual cymes with 1–2 dichasia; peduncle red, 2–6 cm long; bracts persistent and conspicuous, green or pink, obovate, 2.5–12 mm long, acute; pedicels 7–40 mm long. Male flowers: tepals 4, orange or rarely yellow, the outer suborbicular, ovate or elliptic, 7–18 × 6–15 mm, the inner elliptic to spatulate, 6–15 × 4.5–8(–11) mm; stamens 40–60, filaments 0.5–3 mm long; anthers oblong, 0.7–1.8 mm long. Female flowers: tepals 5, orange or yellow, the outer elliptic or ovate, 6–17 mm long, 5–11 mm wide, the innermost elliptic, to 14.5 × 6.2 mm; ovary ellipsoid, 4.5–9 × 2–5 mm, 3-locular, 3-winged; placentation axile, placentas undivided; styles three, caducous, divided about halfway, the branches twisted, with a continuous undulating stigmatic band. Fruit an ellipsoid or cylindrical capsule, 6–28 mm long, 3–7 mm in diameter excluding the wings; wings subequal, triangular, 6–14 mm wide, sometimes extending distally beyond the capsule; dehiscence along the junction with the wing. Fig. 3, p. 16.

TANZANIA. Kigoma District: Kasye Forest, 19 Mar. 1994, *Bidgood, Mbago & Vollesen* 2819!; Iringa District: Mt Image, 1 Mar. 1962, *Polhill & Paulo* 1629!; Rungwe District: Rungwe Forest Reserve, Jan. 1954, *Semsei* 1574!

DISTR. **T** 4, 6–8; Congo-Kinshasa, Zambia, Malawi, Mozambique, Zimbabwe, South Africa

HAB. On tree-trunks or horizontal branches, on wet rock, occasionally terrestrial in moist forest or near streams or waterfalls; 950–2200(–2800) m

CONSERVATION NOTES. Least concern (LC); widespread but uncommon in Flora area

SYN. *B. flava* Marais in Fl. Pl. S. Afr. 31: t. 1233 (1956); Irmsch. in E.J. 81: 172 (1962). Type: Mozambique, Manica, Garuso, 1952, *Schweikerdt* s.n. (PRE, holo.)

B. sutherlandii Hook.f. var. *subcuneata* Irmsch. in E.J. 81: 164 (1962). Type: Tanzania, Rungwe District: Kyimbila, *Stolz* 1042 (B, holo.)

B. sutherlandii Hook.f. subsp. *latior* (Irmsch.) Kupicha in F.Z. 4: 503 (1978). Type: Zambia, above Mukuma, Inono Falls, *Richards* 3699 (B, holo.; K, SRGH, iso.)

B. sutherlandii Hook.f. var. *latior* Irmsch. forma *densiserrata* Irmsch. in E.J. 81: 169, t. 9 fig. 2 (1961). Type: Tanzania, Mbeya District: Utengule, *Stolz* 160 (B, holo.)

B. sutherlandii Hook.f. var. *rubrifolia* Irmsch. in E.J. 81: 169 (1962). Type: Tanzania, Njombe District: Kipengere Range, *Richards* 7767 (K!, holo.)
B. sutherlandii Hook.f. var. *minuscula* Irmsch. in E.J. 81: 170 (1962); Wilczek in F.C.B., Begoniaceae: 51 (1969). Type: Tanzania, Songea District: Matogoro Hills, *Milne-Redhead & Taylor* 8471 (K!, holo.)

NOTE. Irmscher described a considerable number of sub-specific taxa, exhibiting only minor differences. Two subspecies upheld in F.Z. are a good example: subsp. *sutherlandii*: leaves narrowly ovate, 2.5–7 cm long, the margin coarsely serrate; petiole 0.7–2.5 cm long; and subsp. *latior* (Irmsch.) Kupicha: leaves oblong, 5–14 cm long, broadest at the middle, the margin finely serrate; petiole 1–8 cm long. There are numerous intermediates, for example with large leaves and coarsely dentate margins, or with small leaves with rather long petioles. Accordingly, for this Flora, a broader approach is taken and *B. sutherlandii* is seen as a single variable taxon.

The type was said by Irmscher to be from South Africa, Natal, without further locality, Dec. 1861, *Sutherland* s.n. (K, holo.); but the protologue clearly says that the description has been drawn up from a specimen cultivated in UK, York, June 1867, *Backhouse* s.n. (not found)

17. **Begonia tayloriana** *Irmsch.* in E.J. 81: 122 (1961). Type: Tanzania, Ufipa District: Chala Mt, *Richards* 7203 (K!, holo.)

Herb, 10–25 cm high; stem erect, branched, shortly pilose. Stipules persistent, membranous, ovate to oblong, 9–16 mm long, with fimbriate-dentate margin. Leaves asymmetrically ovate, 2.5–4.5 × 2.5–3.5 cm, base slightly cordate, margin with 2(–3) short lobes, otherwise crenulate-serrulate, apex acute, pilose above and on the veins beneath; venation palmate; petiole 1–3.5 cm long. Flowers in a few-flowered cyme; peduncle 8–13 mm long; bracts persistent, oblong, 5–6 mm long; pedicels 10–21 mm long; bracteoles absent. Male flowers: tepals 4, pink, the outer suborbicular, 10 × 10 mm, the inner obovate, 9 × 5–6 mm; stamens many, filaments 1–2 mm long, anthers obovoid. Female flowers: tepals 3, pink, elliptic, 10 × 7 mm; ovary ovoid, 5–6 × 3–5 mm, 3-locular, 3-winged; placentation axile, bilamellate; styles persistent, branched halfway up, the branches twisted, with a spiral stigmatic band. Fruit an ovoid capsule, 9 × 6 mm, one wing larger than the others; wings triangular-deltoid, 5–7 mm wide, rounded; dehiscence along the junction with the wing.

TANZANIA. Ufipa District: Mt Chala, 10 Dec. 1956, *Richards* 7203!
DISTR. **T** 4; only known from the type
HAB. "In a crevice under a rock"; 2100 m
CONSERVATION NOTES. Vulnerable (VU); known only from the type

NOTE. Irmscher just states this is 'well separated from its relatives' by the few triangular 2–3-lobed leaves with acute apex and the short white indument. This comes close to *B. princeae* and is distinct only in the slight leaf lobes and the three (rather than four) ♀ tepals.

18. **Begonia wakefieldii** *Engl.* in Veg. Erde 9: 620 (1921); Irmsch. in E.J. 81: 127 (1961). Type: (not mentioned in protologue) Kenya, 'nyika country', *Wakefield* via *Grant* s.n. (K!, lecto. chosen here)

Erect annual or short-lived perennial herb, terrestrial or on rock, 8–30 cm high, rather succulent, with short creeping rhizome or (fide *Gilbert* 6016) a small subspherical tuber; stems fleshy, pink or green, glabrous. Stipules persistent, triangular, 2–9 mm long, with entire margins. Leaves green on both sides, rarely pink beneath, slightly fleshy, broadly ovate, strongly asymmetric, 4–12 × 2–10.5 cm, base cordate, margins crenate-dentate and often slightly lobed, apex acute to slightly attenuate, glabrous or pilose above; venation ± palmate; petiole green, 1.5–14 cm long. Inflorescence bisexual, in lax terminal cymes of 2–5 flowers; peduncle 1–6 cm long; bracts conspicuous, caducous at length, ovate to obovate, 3–5 mm long, apex rounded; pedicels 5–10 mm long. Male flowers: tepals 4, pale pink (rarely white), the outer suborbicular or broadly elliptic, 5–9

× 4–8 mm, the inner elliptic to spatulate, 4–6 × 2–3 mm long; stamens 28–34, filaments 0.8–2 mm long, anthers ellipsoid, ± 1 mm long. Female flowers: tepals 5, pale pink or rarely white, the outer elliptic, 5–8 × 2–5.5 mm, the inner slightly narrower; ovary ellipsoid, 6–7.5 × 3.5–4 mm, 3-locular, 3-winged; placentation axile; styles three, persistent, divided about halfway, the branches twisted, with a spiral stigmatic band. Fruit an ovoid capsule, 7–13 × 4–5.5 mm excluding the wings; wings unequal, the largest triangular, 7–17 mm wide, the 2 smaller ones to 3 mm wide; dehiscence along the junction with the wing or by breaking up at the loculus wall near the wings.

KENYA. Kilifi District: Rabai, Aug. 1937, *van Someren* 7166!; Kwale District: Shimba Hills, Sheldrick Falls, 23 Oct. 1977, *Gillett & Stearn* 21614! & same loc., 27 Dec. 1980, *Gilbert* 6016!

TANZANIA. Lushoto District: Usambaras, Mt Mlinga, 17 Aug. 1950, *Verdcourt* 306!; Kilosa District: Ruaha River, Kidatu power station, 3 Feb. 1981, *Mhoro* 2754!; Masasi District: Ndanda Mission, 17 Mar. 1991, *Bidgood, Abdallah & Vollesen* 2051!

DISTR. **K** 7; **T** 3, 4, 6–8; not known elsewhere

HAB. Forest or coastal evergreen bushland, rarely miombo woodland; within these vegetation types often on shaded rock or steep banks; 60–1500 m

CONSERVATION NOTES. Data Deficient (DD) as number and size of populations are unknown

SYN. *B. wakefieldii* Engl. forma *dentatilobata* Irmsch. in E.J. 81: 128 (1961). Type: Tanzania, Lushoto District: E Usamabara, 2.4 km N of Pandeni, *Drummond & Hemsley* 3478 (K, holo.), **syn. nov.**

NOTE. The form described by Irmscher is hardly distinct.

19. **Begonia humilis** *Dryand.* in Aiton, Hort. Kew 3: 353 (1787) & in T.L.S. 1: 166, t. 15 (1791). Type: cultivated in Kew from material originating in West Indies (?K, not found)

Terrestrial annual or short-lived perennial herb, 8–20 cm high; stem glabrous. Stipules persistent, ovate, 4–5 × 2.2–2.5 mm, with slightly hairy margins. Leaves asymmetric with the main vein at an angle to the petiole, narrowly ovate, 3–8 × 1–3 cm, one side larger than the other, having a basal lobe with 3–4 main veins, base slightly cordate and very unequal, margin doubly dentate and sometimes slightly lobed, the teeth drawn out into hairs to 1 mm long, apex acute, indumentum restricted to few simple hairs on midrib near base; venation palmate-pinnate; petiole 0.5–2 cm long, with scattered hairs near the junction with the lamina. Inflorescence axillary bisexual cymes of 2–4 flowers; fruiting inflorescences remaining at lower nodes after leaves have fallen; peduncle 21–30 mm long, glabrous; bracts persistent, opposite, ovate, 1–1.5 mm long, with marginal hairs; pedicels 2–6 mm long. Male flowers: tepals 2, white, broadly elliptic, 1.7–2.7 × 3.2 mm; stamens 10–11, arranged on a column 0.7 mm long, filaments 0.2–0.4 mm long, anthers ellipsoid, 0.7–0.8 mm long, the connective slightly extended. Female flowers: tepals 4, white, elliptic, 2 × 0.9–1 mm, subequal; ovary ellipsoid, 2.7–2.8 × 1.7 mm, 3-locular, 3-winged; placentation axile, bilamellate; styles persistent, three, ± 1 mm long, divided near base, the arms twisted with a spiral stigmatic band. Fruit an ovoid to ellipsoid capsule, 6–8.5 × 4–4.5 mm excluding the wings; wings subequal to slightly unequal, 2–4.5 mm wide; dehiscence along the junction with the wing.

TANZANIA. Lushoto District: Amani, May 1916, *Zimmermann* G6651! & "Amani, Muheza District", Nov. 1987, *Kibuwa* 5477!

DISTR. **T** 3; native of the West Indies to Peru and Brazil

HAB. Both times collected as apparently wild, from evergreen forest, wet ground (fide Kibuwa) and in forest in shade (fide Zimmermann); ± 900 m

NOTE. This species occurs quite frequently as a garden escape and has become naturalized in a number of countries outside its native range, like *B. hirtella* Link; the latter has not yet been found in our area.

SPECIES OF DOUBTFUL OCCURRENCE

Begonia horticola *Irmsch.* in E.J. 57(2): 242 (1921); Wilczek in F.C.B., Begoniaceae: 23 (1969); de Wilde in Wageningen Univ. Pap. 2001–2: 113 (2002). Type: Congo-Kinshasa, wild in Botanic Garden of Eala, *Chevalier* 28045 (P, lecto., chosen by de Wilde)

Epiphyte or scrambling terrestrial; stem branched, up to 8 mm wide, rooting at the nodes, becoming woody at the base; indumentum of peltate scales with fimbriate margins. Stipules narrowly triangular, 1–2.4 cm long, persistent, not boat-shaped, keeled, apex acuminate, indumentum present on outside, glabrous within. Leaves thick, papery when dry, ± symmetrical or weakly oblique, elliptic to ovate, 6–14.5 × 3–7 cm, base obtuse to rounded, margin entire to remotely denticulate, apex acuminate, glabrous above, sparsely hairy beneath with dense indumentum on venation; venation pinnate; petiole 1.5–5 cm long, tomentose. Inflorescence axillary, a unisexual dichasial cyme; male inflorescence with up to 15 flowers, peduncle 4–15 mm long, each ramification subtended by a pair of opposite chartaceous, usually connate bracts up to 5 mm long, smaller in the upper parts of the inflorescence, not boat-shaped, apex acuminate; pedicels 0.4–1 cm long; bracteoles absent. Male flowers: tepals 4, free, pink with dark red venation, the outer obovate to elliptic, 0.5–1.1 × 0.3–0.6 cm, the inner narrowly obovate, 0.3–0.7 × 0.1–0.3 cm; stamens 20–40, filaments fused at the base into a column; anthers narrowly ovoid, up to 3 mm long. Female inflorescence with 3 flowers, peduncle ± 1.4 cm long; bracts similar to those in male inflorescence. Female flowers: sessile at first, tepals 4, free, colour and shape as in the male flowers, the outer 8–10 mm long, the inner 5–9 mm; ovary fusiform, without wings, 3–4-locular; placentation septal; styles not persistent in fruit; 3–4, simple, unforked, glabrous, yellow at the base, becoming darker towards the red simple stigmatic apex. Mature fruits with a pedicel up to 1.7 cm long, fleshy, fusiform, 3–5(–6) cm long, terete in cross section, reddish, dehiscent by longitudinal slits.

DISTR. Uganda (no specimens seen; see Note); Congo-Brazzaville, Congo-Kinshasa

SYN. *B. parva* Sprague in Kew Bull. 7: 329 (Sep. 1912). Type: cultivated at Kew, 8 July 1912 (K, holo.), *non* Merrill (Jan. 1912)
B. spraguei C. Weber in Baileya 16, 4: 126 (1968), *nom. nov.* for *B. parva* Sprague

NOTE. de Wilde cites *Purseglove* 2674 from Uganda at K but says he has not seen the specimen. HJB has seen this specimen and thinks it is *B. eminii*. This would mean that *B. horticola* does not occur in East Africa, but could possibly be found there in future. It is however included in the key.

INCERTA SEDIS

K 4; KENYA. Kiambu District: Thika, near Chania Falls, 20 Oct. 1967, *Faden* 67/789! – probably garden escape.

OTHER NAMES

Begonia sultani Hook.f. is a name cited under *B. heddei* by Warburg in Gartenflora 49: 1, fig. 1470 (1900). As far as we know, it has not been published.

Begonia sonderiana Irmsch. is a name often found on labels of specimens from our area; the type is from Mozambique and the species is restricted to that country, Zimbabwe and South Africa.

The following key to Sections is based only on species found in the FTEA area:

1. Fruit fleshy, berry-like 2
   Fruit dry, leathery or papery, regularly dehiscent 5
2. Fruit indehiscent 3
   Fruit dehiscing by longitudinal slits; fimbriate peltate scale present . . . . Section **Tetraphila**
3. Fimbriate peltate scales present 4
   Fimbriate peltate scales absent . . . . Section **Mezierea**
   . . . . 1. *B. meyeri-johannis*; 2. *B. oxyloba*
4. Large bracts enveloping the inflorescence; flowers with 2 tepals fused to form a perianth cylinder . . . Section **Squamibegonia**
   . . . . 3. *B. ampla*; 4. *B. poculifera*
   Inflorescence not enveloped by large bracts; flowers with 4 free petals . . . . Section **Tetraphyla**
   . . . . 5. *B. eminii*; 6. *B. kisuluana*; 7. *B. tatoniana*; 8. *B. zimmermannii* & *B. horticola*
5. Placenta bilamellate in each locule 6
   Placenta entire in each locule . . . . Section **Augustia**
   13. *B. princeae*; 14. *B. riparia*; 15. *B. stolzii*; 16. *B. sutherlandii*; 17. *B. tayloriana*
   . . . . 18. *B.wakefieldii*
6. Petiole without a tuft of hairs at the apex 7
   Petiole with a tuft of hairs at the apex . . . . Section **Rostrobegonia**
   . . . . 9. *B. engleri*; 10. *B. johnstonii*; 11. *B. schliebenii*; 12. *B. wollastonii*
7. Leaf margin variable, never doubly dentate . . . . Section **Augustia**
   13. *B. princeae*; 14. *B. riparia*; 15. *B. stolzii*; 16. *B. sutherlandii*; 17. *B. tayloriana*
   Leaf margin doubly dentate . . . . Section **Doratometra**
   . . . . 19. *B. humilis*

INDEX TO BEGONIACEAE

No new names validated in this part

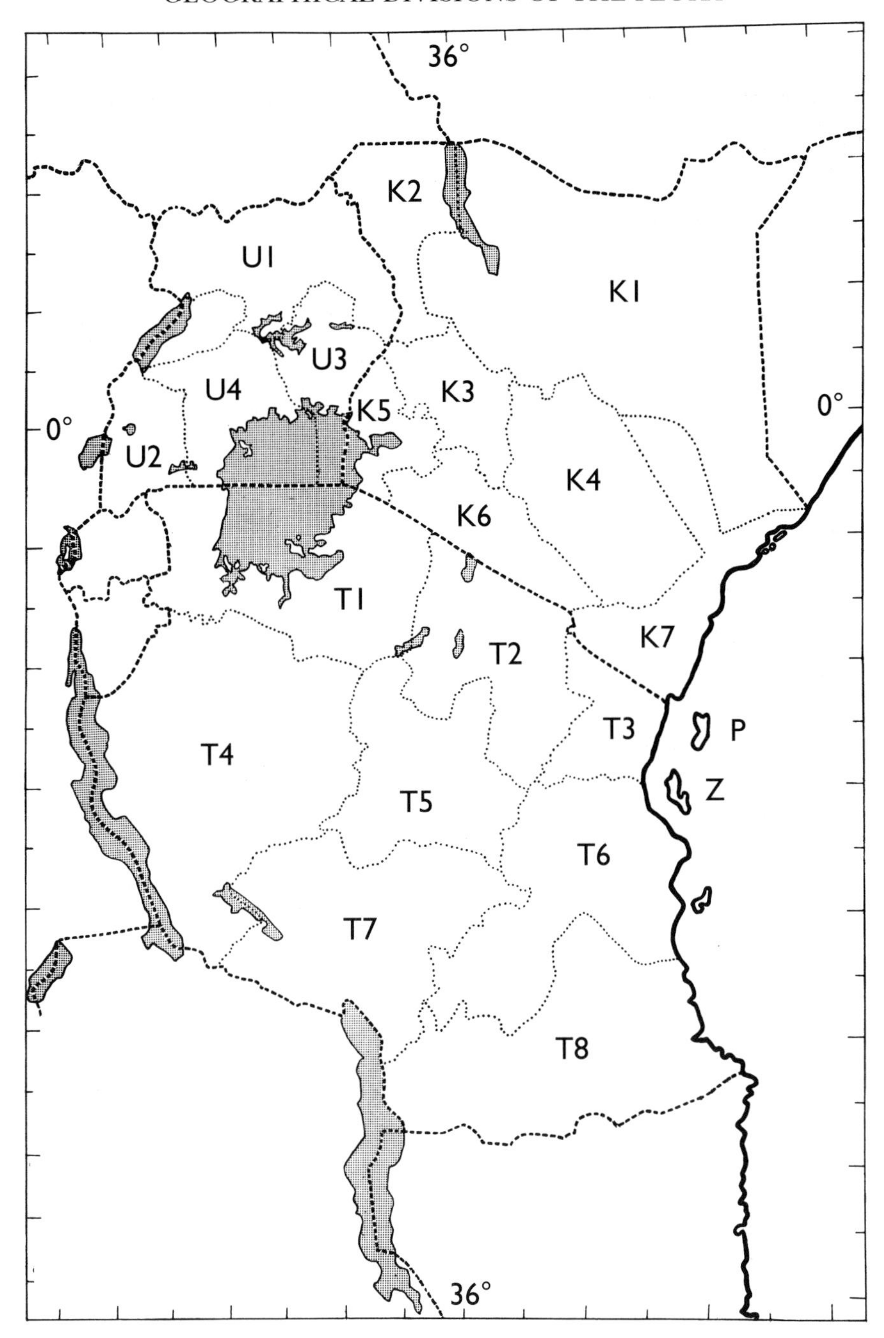
36°
K2
U1
K1
U3
U4
K3
K5
0°
0°
U2
K4
K6
T1
K7
T2
T3
P
T4
Z
T5
T6
T7
T8
36°

First published in 2006 by
Royal Botanic Gardens, Kew
Richmond, Surrey, TW9 3AB, UK
www.kew.org

ISBN 1 84246 161 3

British Library Cataloguing in Publication Data
A catalogue record for this book is available from the British Library

Design and typesetting by Margaret Newman,
Kew Publishing, Royal Botanic Gardens, Kew.

Printed in the UK by Hobbs the Printers

For information or to purchase all Kew titles please visit
www.kewbooks.com or email publishing@kew.org

All proceeds go to support Kew's work in saving the world's plants for life